Emperor Penguin Migration

By Susan H. Gray

21st Century Junior Library

CHERRY LAKE PRESS

Published in the United States of America by Cherry Lake Publishing Group
Ann Arbor, Michigan
www.cherrylakepublishing.com

Reading Adviser: Marla Conn, MS, Ed., Literacy specialist, Read-Ability, Inc.
Content Adviser: Sheila K. Schueller, Ph.D., Lecturer and Academic Program Specialist, University of Michigan School for Environment & Sustainability
Photo Credits: ©robert mcgillivray/Shutterstock.com, cover, 16; ©sirtravelalot/Shutterstock.com, 4; ©David Herraez Calzada/Shutterstock.com, 6; ©Jan Martin Will/Shutterstock.com, 8; ©vladsilver/Shutterstock.com, 10, 14, 18, 20; ©TravelMediaProductions/Shutterstock.com, 12

Cherry Lake Press is an imprint of Cherry Lake Publishing Group.

Library of Congress Cataloging-in-Publication Data

Names: Gray, Susan Heinrichs, author.
Title: Emperor penguin migration / by Susan H. Gray.
Description: Ann Arbor, Michigan: Cherry Lake Publishing, [2021] | Series: Marvelous migrations | Includes index. | Audience: Grades 2-3
Identifiers: LCCN 2020002839 (print) | LCCN 2020002840 (ebook) | ISBN 9781534168589 (hardcover) | ISBN 9781534170261 (paperback) | ISBN 9781534172104 (pdf) | ISBN 9781534173941 (ebook)
Subjects: LCSH: Emperor penguin—Migration—Juvenile literature.
Classification: LCC QL696.S473 G728 2021 (print) | LCC QL696.S473 (ebook) | DDC 598.47156/8—dc23
LC record available at https://lccn.loc.gov/2020002839
LC ebook record available at https://lccn.loc.gov/2020002840

Cherry Lake Publishing Group would like to acknowledge the work of the Partnership for 21st Century Learning, a Network of Battelle for Kids. Please visit http://www.battelleforkids.org/networks/p21 for more information.

Printed in the United States of America
Corporate Graphics

CONTENTS

In Antarctica, temperatures can drop as low as -76°F (-60°C).

Eating and Eating

What a great day in **Antarctica**! The sun has been shining brightly for 24 hours straight. The temperature shot up all the way to 14 degrees Fahrenheit (–10 degrees Celsius). Emperor penguins are having the time of their lives. They're swimming, catching fish, diving, and catching more fish. They have really put on weight lately. But soon enough, these birds will be

Emperor penguins can hold their breath for up to 22 minutes!

fasting. It's late February, and **migration** will begin soon.

More than 500,000 emperor penguins live in the **continent** of Antarctica. In December, January, and February, they live near the coastline. There, they can catch plenty of fish, **crustaceans**, and squid. But in March, they begin to migrate inland.

Think!

These penguins eat something called the giant warty squid. What qualities would a penguin need to catch such food underwater?

Adult emperor penguins have a black head with yellow markings and a white stomach.

The Trek

Every year, emperor penguins must travel to an area where they can breed and lay their eggs in safety. They head south, toward towering ice cliffs. The cliffs offer some protection from the biting winds.

Around early March, fattened penguins begin the trek. Progress is slow. They are lugging a lot of extra weight. At the start, the birds cover just 2 to 3 miles (3 to 5 kilometers) each day.

On their journey, winds can reach up to 90 miles (145 km) per hour.

With their short legs, penguins can take only baby steps. Sometimes, they travel by **tobogganing**. They drop to their bellies, dig their claws into the ice, and shove forward.

The penguins' migration is less than 125 miles (200 km). The trip does not even take a month. Still, nothing is easy about it.

Make a Guess!

Why do emperor penguins burn more energy at the start of their trek? Explain your answer.

Emperor penguins partner with one mate for a season. A season is about 1 year.

Eggs and Babies

Once they arrive at their **breeding grounds**, males begin **courting**. They bow their heads, call, and strut before the females. Several weeks after a pair mates, the female lays one large egg. Just by forming the egg, she has burned extra energy. Now, she really needs to eat. The mother bird moves the egg to her mate. Then, she turns and heads back to the sea.

Even baby chicks still snuggle with their father for warmth.

Females are gone for about 2 months. During that time, they retrace their steps to the coast. They fill up on seafood, gain weight, and hike back.

Meanwhile, each male protects and warms its precious egg. The egg rests on its father's feet. A big fold of feathered skin on his belly covers the egg and keeps it warm. After 2 months, the egg begins to hatch.

Look!

Find some pictures or videos of male penguins huddling close together. Why do you think they do this?

The mother emperor penguin will feed her chick by regurgitating into its mouth.

Mother penguins have perfect timing. Around hatching time, they return with food for their little ones. By now, it's July and their mates have not eaten for 4 months. So it's the males' turn to take off for the coast.

Between now and December, parents will rotate feeding duties.

Ask Questions!

The protective fold of skin on the father's belly is called the brood pouch. Do mother penguins have this too?

Emperor penguin chicks are typically gray with black heads.

Return Trip

As the year comes to a close, temperatures rise. In the warmer weather, ice at the edge of the continent melts. The distance from the breeding grounds to the ocean shortens.

By December, the youngsters have grown quite large. Their parents have fed them well. Chicks can now manage on their own. The adults head north toward the ocean. Chicks

Emperor penguins need just the right amount of ice to safely migrate. Climate change is already putting this species at risk.

follow soon after. A wealth of food awaits back at the coast.

Emperor penguins now have one thing in mind—a seafood feast! The birds waddle and toboggan their way to the coast. For the chicks, it is their first migration adventure. With luck, they will make the round trip for many years to come.

GLOSSARY

Antarctica (ant-ARK-tih-kuh) the landmass around the South Pole

breeding grounds (BREED-ing GROUNDZ) areas where animals go to mate

continent (KAHN-tuh-nuhnt) one of Earth's great landmasses

courting (KORT-ing) behaving a certain way to attract a mate

crustaceans (kruh-STAY-shuhnz) animals such as crabs and lobsters that have an outer shell and no backbone

fasting (FAST-ing) going without food

migration (mye-GRAY-shuhn) movement of animals from one region to another and back again

tobogganing (tuh-BAH-guhn-ing) sliding as if on a sled

FIND OUT MORE

BOOKS

Esbaum, Jill. *Penguins*. Washington, DC: National Geographic Children's Books, 2014.

Marsico, Katie. *Emperor Penguins*. Portsmouth, NH: Heinemann, 2011.

Salomon, David. *Penguins!* New York, NY: Random House Books for Young Readers, 2017.

WEBSITES

Enchanted Learning—Penguins

https://www.enchantedlearning.com/subjects/birds/printouts/penguins.shtml
Learn about some of the different kinds of penguins, including emperors.

National Geographic Kids

https://www.natgeokids.com/uk/discover/animals/birds/emperor-penguins
Read 10 fun facts about the world's largest penguins.

INDEX

ABOUT THE AUTHOR

Susan H. Gray has a master's degree in zoology. She has written more than 170 reference books for children and especially loves writing about animals. Susan lives in Cabot, Arkansas, with her husband, Michael, and many pets.